Engineering for Inclusion 2022

Held at the 2022 AIChE Annual Meeting

Phoenix, Arizona, USA
13-18 November 2022

ISBN: 978-1-7138-7908-4

Printed from e-media with permission by:

Curran Associates, Inc.
57 Morehouse Lane
Red Hook, NY 12571

Some format issues inherent in the e-media version may also appear in this print version.

Printed with permission by Curran Associates, Inc. (2023)

For permission requests, please contact AIChE
at the address below.

AIChE
120 Wall Street, FL 23
New York, NY 10005-4020

Phone: (800) 242-4363
Fax: (203) 775-5177

www.aiche.org

Additional copies of this publication are available from:

Curran Associates, Inc.
57 Morehouse Lane
Red Hook, NY 12571 USA
Phone: 845-758-0400
Fax: 845-758-2633
Email: curran@proceedings.com
Web: www.proceedings.com

TABLE OF CONTENTS

MAC EMINENT ENGINEERS AWARDS POSTER SESSION

225a Water Orientation and Dynamics in Cavitand Pockets 1
Busayo Alagbe, Bruce C. Gibb, Henry Ashbaugh

225b Computational Studies of the Order-Disorder Transition in Block Copolymer Topological Blends 2
Rahul Kumar, Amy Goodson, Oluwafemi Alli, Clayton Chamness, Isabella Miserocchi, Julie Albert, Henry Ashbaugh

Incorporating Multiple Sites in the Design of Alloy Catalysts for Nitrogen Activation and Spillover........ 3
Chukwudi Nwaokorie

225c Non-Isothermal Crystallization Kinetics of Poly (ε-caprolactone) (PCL) and MgO Incorporated PCL Nanofibers........ 4
Daisaku Gicheha, Nabila Shamim

225d Bench-Scale Testing of Electrochemical Recovery of Phosphorus from Post-Digester Municipal Wastewater Driven By Magnesium Salt - MAC Poster Session 5
Lawrence Ajayi, Sana Heydarian, Jason Trembly, Damilola Daramola

225e Evaluating Transport Factors to Understand Electrochemical Nutrient Removal and Recovery from Synthetic Animal Wastewater- MAC Poster Session........ 7
Sana Heydarian, Francesca Carney, Damilola Daramola

225f Transfer Learning Framework for Catalysis........ 9
Gbolade Kayode

225g Conceptual Design of Nano-Straws in Spherical Porous Particles for Improved Adsorption Capacity for Carbon Capture........ 10
Oluwole Ajumobi

225h Evaluating the Role of Carbon Microspheres in an Aqueous Lubricant - The Stribeck Curve Analysis........ 11
Samuel Solomon, Noshir Pesika

225i Bioconversion of Plastic Wastes into Value-Added Products Using Thermal Oxo-Degradation 12
Efrain Rodriguez-Ocasio, Mark Blenner, Laura R. Jarboe

Author Index

225a Water Orientation and Dynamics in Cavitand Pockets

Busayo Alagbe, Bruce C. Gibb, Henry Ashbaugh

Hydrophobic interactions drive the binding of nonpolar ligands to the oily pockets of proteins and supramolecular species in aqueous solution. Thus, the wetting of host pockets is expected to play a critical role in determining the thermodynamics of guest binding. This however requires a good understanding of the properties of water molecules in such hosts since they can be very different from those in the bulk. This is expected to unveil the role of water in guiding solvent specific interactions between host and guest species in solution. In this work, we use molecular simulations to examine the structure of water in the pockets of deep-cavity cavitands and the kinetics of their filling and emptying. Deep-cavity cavitands are a class of water-soluble, bowl-shaped supramolecular host species possessing a nonpolar guest binding pocket that is approximately 8 Å deep with a portal opening approximately 8 Å wide. Here, we employed the non-functionalized parent cavitand *octa-acid*, *tetra-endo-methyl octa-acid* (which differs from the parent by four methyl groups oriented inward towards the pocket) and *positand* (which differs from the parent by replacing the acid groups with quaternary ammoniums). Based on the water occupation state of each cavity, we mapped out via graph analysis the different water structures using hydrogen bonding as connection between nodes. This enabled us to distinguish the water distribution and orientation in the different cavitands as well as to determine the regimes and stability of the structures. Our results also show that empty pockets are especially long lived compared to their filled counterparts which suggests that the emptying of hydrophobic pockets will play a key role in binding kinetics.

225b Computational Studies of the Order-Disorder Transition in Block Copolymer Topological Blends

Rahul Kumar, Amy Goodson, Oluwafemi Alli, Clayton Chamness, Isabella Miserocchi, Julie Albert, Henry Ashbaugh

Block copolymer (BCP) based nanolithography provides an economical way to pattern nanoscale features for the semiconductor industry. Cyclization can reduce feature sizes, but like linear BCPs, cyclic BCP nanostructure size is still limited by the order-disorder transition (ODT). Although the ODTs of pure linear and cyclic BCPs have been studied extensively, there is little information on binary blends of these BCPs. We use dissipative particle dynamic simulations to study the impact of size mismatch and molecular architecture of component BCPs on the ODT for various blends. We see that the blend ODT always occurs at higher segregation strength than one would predict from linear interpolation of pure component ODTs. The deviation from this simple prediction is greater for blends with greater size mismatch between components. We find clustering of like components (linear with linear, cyclic with cyclic) in the disordered phase. Further, there exists a significant mismatch between the characteristic lengths associated with the component structures at the predicted segregation strength for blends with greater size mismatch between components. This results into the requirement of higher segregation strength to facilitate the cluster formations of comparable characteristic lengths for the transition to occur.

Incorporating Multiple Sites in the Design of Alloy Catalysts for Nitrogen Activation and Spillover

Chukwudi Nwaokorie

Ammonia (NH_3) is one of the most important feedstocks for the production of chemicals and fertilizer making it vital to the global economy at large. Activation of molecular nitrogen (N_2) is seen as a critical step in the Haber-Bosch process, which is the ubiquitous industrial process for the production of ammonia. It has been shown that the activity of mono-metallic catalysts commonly used for ammonia synthesis is restricted by the scaling relation that low activation energy of N_2 correlates with strong adsorption of intermediates, which blocks adsorption sites and hinders NH_3 synthesis. Heterogeneous catalysts that allow facile activation of N_2 and weak binding of N atoms would exhibit the optimal efficiency for these reactions. In this study, we employ first principles calculations and micro-kinetic modeling to study N_2 activation and spillover on alloy surfaces. We demonstrate that incorporating single-atom alloy sites allows breaking scaling relations and fosters N spillover. This could lead to the development of new catalysts with enhanced activity due to the low N_2 activation energy barrier relatively weakly bound N.

225c Non-Isothermal Crystallization Kinetics of Poly (ε-caprolactone) (PCL) and MgO Incorporated PCL Nanofibers

Daisaku Gicheha, Nabila Shamim

Background: An understanding of the kinetics of the crystallization process is important for the selection of processing parameters such as temperature and time during the fabrication process. Differential scanning calorimetry (DSC) is an excellent tool for following the progress of crystallization. **Objective**: the objective of this study is to investigate the non-isothermal crystallization kinetic of pure polycaprolactone (PCL), and magnesium Oxide nanoparticle incorporated PCL (MgO-PCL) nanofiber composite. **Materials and Methods**: PCL and MgO-PCL nanofiber composite are analyzed by Fourier-transform infrared spectroscopy (FTIR), differential scanning calorimetry (DSC), and X-ray diffractometry (XRD). **Results:** The investigation was made as a function of cooling rates from 0.5 – 30 K/min to observe the crystallization, and melting behavior of poly (ε-caprolactone) and MgO-PCL nanofibers using DSC-3 from Mettler Toledo. The result shows that the crystallization temperature increased as the cooling rate decreased but with little melting temperature dependence on the cooling rates. Furthermore, the nanofibers' structural analysis and chemical structure were investigated using XRD and FTIR, respectively. The change in crystallinity using XRD indicates a change in chain mobility and might translate to higher mechanical properties. **Conclusions:** Polymeric electro spun nanofibers reinforced with nanoparticles (NPs) present many potential applications since they combine the flexibility and porosity of electro spun polymeric fibers with the functional properties of Mg nanoparticles.

225d Bench-Scale Testing of Electrochemical Recovery of Phosphorus from Post-Digester Municipal Wastewater Driven By Magnesium Salt - MAC Poster Session

Lawrence Ajayi, Sana Heydarian, Jason Trembly, Damilola Daramola

Phosphorus is an irreplaceable macronutrient necessary for plant growth and other industrial processes, whereas when in excess in water bodies can lead to eutrophication which causes algal blooms, reduced water clarity, and toxicity to aquatic life due to lack of oxygen. Through human excrements, dishwashing agents, and other unspecific sources, approximately 1.8 g phosphorus per capita per day reaches the wastewater treatment plants [1]. This wastewater, low in phosphorus concentration, passes through a primary clarifier and then a digestion process. The typical digestion process aims to reduce the amount of sludge that needs to be discarded, in this process, a phosphorus-rich supernatant develops. In addition, as demand for fertilizers increases globally and the supply of non-renewable rock phosphate steadily decreases, the costs of phosphate-based fertilizers will increase, significantly impacting the affordability of food products [2, 3]. Recovery of phosphorus from waste, as struvite ($MgNH_4PO_4 \cdot 6H_2O$): a slow-release fertilizer, is a potential pathway to solving these problems. Therefore, it is imperative to develop a technology to optimize phosphorus recovery from post-digester wastewater as this could lead towards environmental sustainability and create economic value.

This research study focuses on recovering phosphorus as struvite (or a mixed-phase) from post-digester municipal wastewater using electrochemical technology. The electrochemical technique adopted provides localized pH control which is necessary for struvite formation through oxygen reduction at the cathode. This approach also aims to achieve a procedure that enables easy automation with little or no human intervention. Belarbi *et al.* investigated the impact of the presence of other dissolved species specifically Ca^{2+} in wastewater and how it can alter the chemistry of struvite precipitation [4]. This provided insight into negating the effect of this competing ion by increasing the concentration of magnesium in the wastewater in addition to other major ions in this stream including Na^+, K^+, SO_4^{2-} and Cl^-. In a bid to potentially recover phosphorus as struvite in this study, magnesium will be supplemented by the introduction of magnesium salt.

The effect of the available surface area of the working electrode, flow turbulence, applied potential, and dissolved oxygen present will be investigated. Also, the percentage recovery of phosphorus as struvite will be determined while analyzing it using various characterization techniques such as scanning electron microscope, Energy-dispersive X-ray spectroscopy, Raman spectroscopy, and X-Ray Diffraction.

References:

[1] Cornel, P., andamp; Schaum, C. (2009). “Phosphorus recovery from wastewater: needs, technologies and costs” *Water Science and Technology*, 59(6), 1069–1076, 2009.

[2] László Kékedy-Nagy, Leah English, Zahra Anari, Mojtaba Abolhassani, Bruno G. Pollet, Jennie Popp, Lauren F. Greenlee, “Electrochemical nutrient removal from natural wastewater sources and its impact on water quality” *Water Research,* 210, 2022.

[3] Arseto Yekti Bagastyo, Anita Dwi Anggrainy, Khoiruddin Khoiruddin, Riang Ursada, IDAA

Warmadewanthi, I Gede Wenten, "Electrochemically-driven struvite recovery: Prospect and challenges for the application of magnesium sacrificial anode" *Separation and Purification Technology*, vol. 288, 2022.

[4] Belarbi, D. Daramola and J. P. Trembly, "*Bench-Scale Demonstration and Thermodynamic Simulations of Electrochemical Nutrient Reduction in Wastewater via Recovery as Struvite*" J. Electrochem. Soc., vol. 167, no. 4, Nov. 2020.

225e Evaluating Transport Factors to Understand Electrochemical Nutrient Removal and Recovery from Synthetic Animal Wastewater- MAC Poster Session

Sana Heydarian, Francesca Carney, Damilola Daramola

Background

Concentrated Animal Feeding Operations (CAFOs) are one of the main sources of introducing excessive nutrients, mainly phosphorus and nitrogen, to the environment via manure use as fertilizer. [1] These excessive nutrients are key factors in eutrophication of water bodies, and it is crucial to treat the CAFO effluents before disposal to water bodies, in order to reduce environmental impacts. [2] Furthermore, limited natural resources of phosphorus increases the need for an alternative resource for this nutrient. Simultaneous removal and recovery of phosphorus and nitrogen from wastewaters via an electrochemical approach, as a precipitate called $MgNH_4PO_4{\cdot}6H_2O$, i.e., struvite, addresses both excessive amount of nutrients in these effluents and limited resources of natural phosphorus. Struvite also has economic benefit as a slow-release fertilizer. [3] A predictive model of the effects of chemical composition and solution conditions on phosphorus removal from wastewater and recovery as struvite has been introduced in literature. [4] This model requires additional experimental data to validate (or modify) the anticipated effects of temperature and ion concentration on solid phosphorus recovery.

Methodology

In the current study, the effect of ion concentration, temperature, solution agitation, and pH on nutrient removal efficiency and the type of P-containing precipitant recovered will be investigated. This investigation will occur in a bench scale setup with well-defined mass and thermal transport. This study will help with (1) providing a fundamental background knowledge for electrochemical nutrient recovery and removal, (2) enhancing the actual efficiency of electrochemical nutrient recovery by separating the recovery due solely to chemical transport and (3) assisting with validating/modifying the aforementioned literature model.

Results

The initial experiments of this study showed higher phosphorus removal from wastewater in pH of 9 in comparison to pH of 7.6, in agreement with the model prediction. [4] Further investigation of the collected solid phase samples will be used to assess the effect of pH on the type of recovered precipitate. Additional results of solid and liquid phase analyses will be presented at the conference including the effect of other transport factors on the recovered nutrient and the delineation between literature model prediction and experimental results.

References

[1] Z. Belarbi, D. A. Daramola and J. P. Trembly, "Bench-Scale Demonstration and Thermodynamic Simulations of Electrochemical Nutrient Reduction in Wastewater via Recovery as Struvite," J. Electrochem. Soc., vol. 167, no. 4, Nov. 2020, doi: 10.1149/1945-7111/abc58f.

[2] L. Kekedy-Nagy*, et. Al*, "Electrochemical Nutrient Removal from Natural Wastewater Sources and its Impact on Water Quality,"

[3] Z. Belarbi and J. P. Trembly, "Electrochemical Processing to Capture Phosphorus from Simulated Concentrated Animal Feeding Operations Waste," J. Electrochem. Soc., vol. 165, no. 13, pp. E685–E693, Jan. 2018, doi: 10.1149/2.0891813jes.

[4] G. P. Pindine, J. P. Trembly and D. A. Daramola, "Equilibrium-Based Temperature-Dependent Economic Analysis of the Recovery of Phosphorus from Different Wastewater Streams via Chemical Precipitation" ACS EST Water, 2021, **1**, 2318–2326.

225f Transfer Learning Framework for Catalysis

Gbolade Kayode

Machine learning has been successfully applied in recent years to screen materials for a variety of purposes. However, despite recent advances, most screening-based machine learning approaches are limited in generality and transferability, requiring new models to be created from scratch for each new application. This is particularly apparent in catalysis, where there are many possible intermediates and transition states of interest, in addition to a large number of potential catalytic materials. In this work we developed a new machine learning architecture which is built on chemical principles and allows creation of general, reusable models. Our new architecture uses latent variables to simplify the learning process, thereby making it more data-efficient while promoting transfer learning. We show that this architecture allows the creation of models that can be reused for many different applications, providing improved efficiency and convenience. The integration of latent variables also provides physical interpretability, as predictions can be explained in terms of the learned chemical environment as represented by the latent space. Lastly, we show that our new machine learning architecture is general and robust enough to handle multiple types of systems.

225g Conceptual Design of Nano-Straws in Spherical Porous Particles for Improved Adsorption Capacity for Carbon Capture

Oluwole Ajumobi

Amine functionalization of high surface area MCM-41 particles for CO_2 capture, using polymeric amines, is well established but has a drawback of the timely degradation and leeching of the polymer when exposed to moisture or wet environment. Alternatively, chemical (covalent) grafting amine functional groups from an alkoxysilane such as 3-Aminopropyl triethoxysilane (APS) on MCM-41 offers better stability against these drawbacks. However, the diffusional restriction exhibited by the narrow uniform MCM-41 pores (2 – 4 nm) impedes the effective amine functionalization of all available silanol groups within the inner mesoporous core. This leads to incomplete amine functionalization and thus, the CO_2 adsorption capacity is low. To solve this problem, we have integrated 40 wt% of halloysite nanotubes (HNTs) as nano-straws into the spherical MCM-41 spheres (M40H) using a one-step facile aerosol-assisted synthesis technique. This ship-in-a-bottle approach is such that, as the silica growth occurs in the heating zone due to evaporation induced hydrolysis and condensation, and the HNTs are integrated into the mesoporous spherical matrix with a considerable part of their length protruding as straws. Afterwards, the particles are functionalized with varying volume of APS in Toluene at 80 °C for 24 h. Characterization results shows that the structural properties of MCM-41 remained unaltered after the incorporation of the nano-straws and amine functionalization. We showed the CO_2 capture at atmospheric pressure and 35 °C at dry conditions, using a thermogravimetric analysis setup. The results showed that the APS functionalized M40H (APS/M40H_2) has a higher adsorption capacity of 1.81 mmol/g which is 18.3% higher than that of APS functionalized MCM-41 (APS/MCM-41_2). When normalized based on 60 wt% of MCM-41 in M40H, the APS/M40H_2 has an adsorption capacity of 3.02 mmol/g which doubles that of APS/MCM-41_2 (1.53 mmol/g). This shows that the incorporation of HNT as nano-straws leads to enhanced CO_2 capture due to (1) improved accessibility (and diffusion) of APS into the inner core of MCM-41 spheres for enhanced amine functionalization by covalent grafting and (2) enhanced CO_2 capture via improved molecular transport of CO_2 molecules to access the amine sites within the MCM-41 framework. This unique composite morphology of nano-straws in spherical porous particles has potential for other applications to yield similar effect.

225h Evaluating the Role of Carbon Microspheres in an Aqueous Lubricant - the Stribeck Curve Analysis

Samuel Solomon, Noshir Pesika

Carbon Microspheres (CMS) have great potential as additives for enhancing the performance of base lubricants. They act as ball bearings that convert sliding friction into rolling friction, thus reducing the effective friction coefficient. The Stribeck curve is a fundamental tool used in tribology to evaluate the performance of a lubricant. This study aims to investigate the influence that CMS have on the three lubrication regimes of the Stribeck curve - boundary, mixed and hydrodynamic. First, monodispersed CMS with an average size of about 200 nm were synthesized via a hydrothermal process from glucose solution. A base lubricant was prepared by mixing 60 wt% glycerol-water solution with a small amount of sodium dodecyl sulfate surfactant before dispersing varying amounts of CMS to obtain our lubricant-CMS formulation. Friction experiments were conducted using a pin-on-disc tribometer under varying loads and concentrations of CMS in base lubricant. The resulting Stribeck plots reveal that the particles are most effective in the boundary lubrication regime. This reduction in friction coefficient is due to the rolling mechanism of the particles within the contact region of the tribo-pair. However, the results also indicate that the ability of the particles to function as rolling elements between the friction pair diminishes as the film thickness increases in the mixed and hydrodynamic lubrication regimes. This fundamental study provides further insight into the role of microparticles in lubrication.

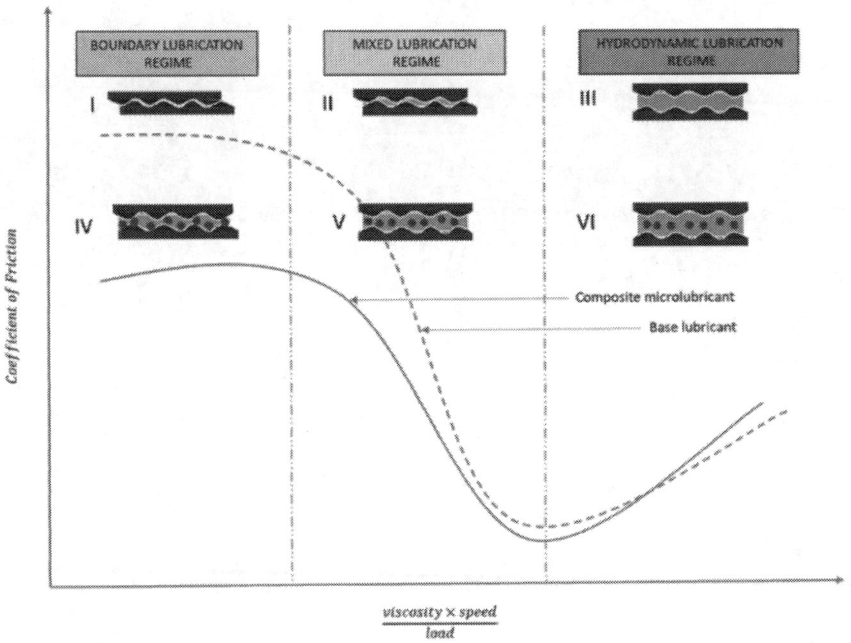

225i Bioconversion of Plastic Wastes into Value-Added Products Using Thermal Oxo-Degradation

Efrain Rodriguez-Ocasio, Mark Blenner, Laura R. Jarboe

Over 8 billion tons of plastics have been produced, and over half of it has ended or will end in landfills or the environment. Plastic accumulates in the environment because evolution has not had enough time to develop a mechanism to break down plastics into molecules that can be recycled by nature. Instead, it can take a thousand years for plastics to break down via abiotic action, making it the rate-limiting step in plastic biodegradation. Thermal oxo-degradation (TOD) can accelerate this rate-limiting step by rapidly deconstructing the polymer backbone and inserting oxygen functionalities to produce lipids and hydrocarbons. At this point, engineered microbes can be employed to convert the carbon that originated from plastic wastes into higher-value products to promote a circular economy and prevent plastic pollution. In this work, non-conventional yeasts were selected against model compounds representative of the composition of thermally oxo-degraded High-Density Polyethylene (HDPE), the most common type of plastic. The best-performing microorganism was able to use TOD products from plastics as the sole carbon source without emulsification, overcoming the mass transfer limitations presented by a hydrophobic and solid substrate. The already remarkable growth profile of the microorganism was improved via adaptative laboratory evolution, increasing the growth rate by >200%. This work represents the proof-of-concept for a promising plastic waste upcycling approach and presents a non-conventional yeast with the potential to become a microbial cell factory capable of using heterogeneous and hydrophobic carbon sources as raw material.

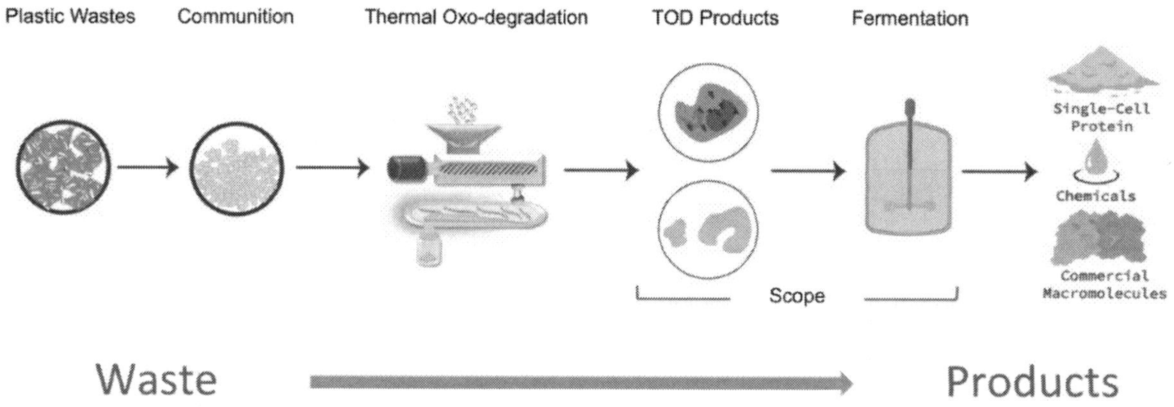

AUTHOR INDEX

Ajayi, Lawrence..5
Ajumobi, Oluwole 10
Alagbe, Busayo..1
Albert, Julie 2
Alli, Oluwafemi 2
Ashbaugh, Henry 1, 2
Blenner, Mark..12
Carney, Francesca 7
Chamness, Clayton 2
Daramola, Damilola...5, 7
Gibb, Bruce C. 1
Gicheha, Daisaku...4
Goodson, Amy 2
Heydarian, Sana...5, 7
Jarboe, Laura R. 12
Kayode, Gbolade 9
Kumar, Rahul...2
Miserocchi, Isabella 2
Nwaokorie, Chukwudi 3
Pesika, Noshir..11
Rodriguez-Ocasio, Efrain 12
Shamim, Nabila 4
Solomon, Samuel...11
Trembly, Jason 5